I0483588

BLAST OFF!

The Physics of **Space**

Written by Arnold Ringstad

WORLD BOOK

www.worldbook.com

Co-published by agreement between Shi Tu Hui and World Book, Inc.

Shi Tu Hui
Room 1807, Block 1,
#3 West Dawang Road
Chaoyang District, Beijing 100025
P.R. China

World Book, Inc.
180 North LaSalle Street
Suite 900
Chicago, Illinois 60601
USA

Library of Congress Control Number: 2024947050

Aha! Academy: Physics
ISBN: 978-0-7166-7144-2 (set, hard cover)

Blast Off! The Physics of Space
ISBN: 978-0-7166-7145-9 (hard cover)
ISBN: 978-0-7166-7165-7 (e-book)
ISBN: 978-0-7166-7155-8 (soft cover)

Printed in India by Replika Press PVT LTD, Haryana, India
1st printing January 2025

Staff

Editorial

Vice President
Tom Evans

Editorial Project Coordinator
Kaile Kilner

Senior Curriculum Designer
Caroline Davidson

Proofreader
Nathalie Strassheim

Graphics and Design

Senior Visual
Communications Designer
Melanie Bender

Designer
Shannon Hagman

Digital Asset Specialist
Rosalia Bledsoe

Written by Arnold Ringstad
Advised by Eric Hazlett

Developed with World Book by
Red Line Editorial

Acknowledgments

The publishers gratefully acknowledge the following sources for photography. All illustrations were prepared by WORLD BOOK unless otherwise noted.

Cover: NASA; NASA, ESA, CSA, STScI, T. Temim (Princeton University); E2 art lab/Shutterstock; Frame Stock Footage/Shutterstock; An Wenhong, Shutterstock

Addictive Stock Creatives/Alamy Images 35; KPNO/NOIRLab/NSF/AURA/T. Slovinský (licensed under CC BY 4.0) 4; Library of Congress 24; NASA 4, 5, 7, 9, 11, 12, 13, 15, 16, 17, 18, 19, 21, 22, 23, 27, 28, 29, 30, 31, 32, 34, 35, 36, 38, 39, 40, 41, 42, 43, 46, 47, 48; Public Domain (Resonance cascade) 26; Shutterstock 4, 5, 6, 7, 8, 9, 10, 11, 12, 14, 15, 18, 19, 20, 21, 23, 24, 25, 26, 27, 29, 31, 33, 34, 35, 36, 37, 40, 41, 42, 43, 44, 45, 46, 47; Matthew Simantov (licensed under CC BY 2.0) 19

There is a glossary of terms on page 48. Terms defined in the glossary are in type that looks like *this* on their first appearance on any spread (two facing pages).

Contents

Introduction

Moons circle their planets in an elegant dance. Asteroids hurtle across the solar system at breathtaking speeds. And mysterious black holes lurk in total darkness, daring us to discover their secrets. Space is a place of beauty, danger, and mystery.

One day, humans may leave our home planet. We will travel to distant star systems, searching for answers to the universe's greatest questions. But until then, we have a set of tools to help us scratch away at the deepest secrets of the cosmos.

Physics is the science behind so many amazing space phenomena. From gravity to atomic *forces* to rocket science, physics is the key to unlocking our universe.

Could humans ever travel to faraway planets? Physics will help us find out!

A UNIVERSAL FORCE

Gravity is the *force* of attraction that draws all *matter* together. Every object in the universe has gravity, from the most massive galaxy to the tiniest speck of space dust. Even you exert your own gravitational pull! The greater an object's mass, the stronger its gravitational field. Earth has more mass than you. This is why objects fall toward our planet instead of falling toward you!

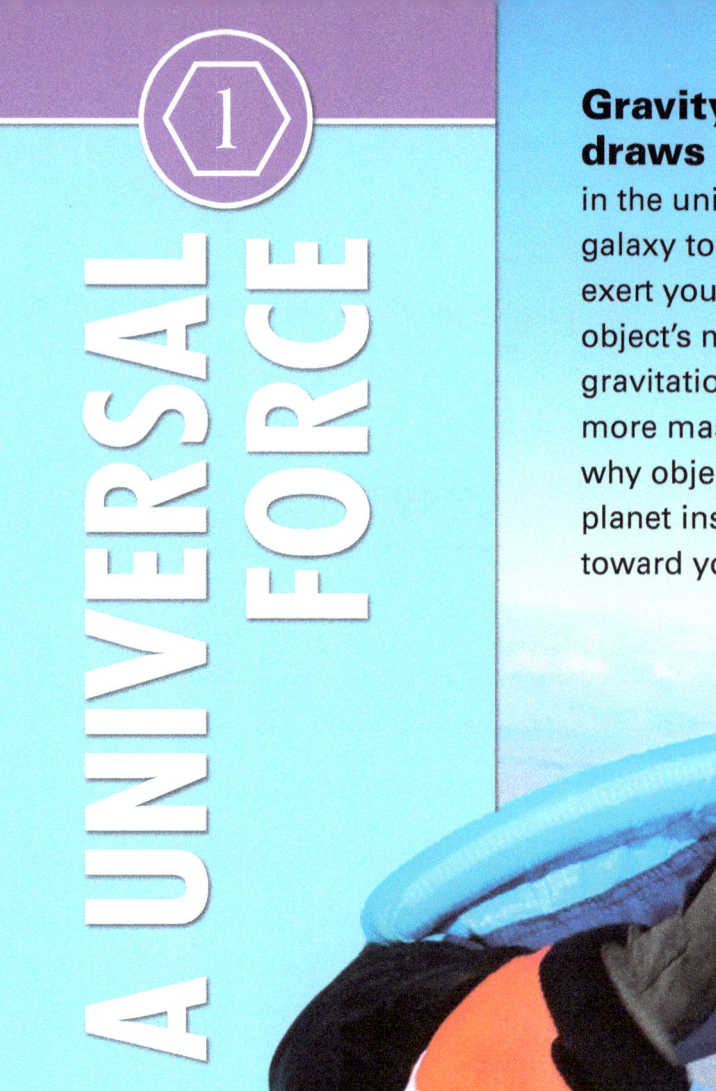

Thanks for the ride, gravity!

An apple falls from a tree. A skydiver plunges toward Earth. A meteor screams through the atmosphere in a dazzling flash of light. Gravity is at work.

The closer two objects get to one another, the stronger their gravitational attraction. So Earth's gravitational pull on you when you're standing in a deep valley would be (slightly!) stronger than if you were standing on a mountaintop.

Gravity can cause two galaxies to collide.

Moon basketball

Space dunk!

On planet Earth, the average professional basketball player can jump more than 2 feet (61 centimeters).
But what if you played a game on the moon? A moon basketball player could jump 14 feet (4.3 meters), clearing the top of the basketball hoop! A basketball player has the same mass no matter where they are in space. But the player's weight depends on the gravitational pull of the celestial body on which they are standing.

Objects with greater mass have stronger gravitational pulls.
Surface gravity is the amount of gravitational pull an object experiences on a planet's or moon's surface. Earth's surface gravity is 1 g.

On a celestial body, such as the moon, gravity pulls objects toward the body's center.

The *force* of this pull is called weight.

	Earth	
CELESTIAL BODY TYPE Planet	**WEIGHT OF 100-KG BASKETBALL PLAYER** 220 pounds (99.8 kg)	
GRAVITY 1 g	**VERTICAL JUMP HEIGHT** 2.3 feet (70.1 cm)	

	Pluto	
CELESTIAL BODY TYPE *Dwarf planet*	**WEIGHT OF 100-KG BASKETBALL PLAYER** 15 pounds (6.8 kg)	
GRAVITY 0.07 g	**VERTICAL JUMP HEIGHT** 36 feet (11 m)	

	Deimos	
CELESTIAL BODY TYPE Martian moon	**WEIGHT OF 100-KG BASKETBALL PLAYER** 0.07 pounds (0.03 kg)	
GRAVITY 0.00031 g	**VERTICAL JUMP HEIGHT** 7,521 feet (2,292 m)	

* The basketball player's mass is 100 kg.

DID YOU KNOW?

Escape velocity is the speed an object must reach to escape a celestial body's gravitational pull. The escape velocity of Deimos is only about 12 miles (20 kilometers) per hour. A person standing on Deimos could probably jump into space!

9

Slingshots in **space**

In a gravity assist, a spacecraft flies near a large celestial body, such as a planet. The planet's gravity draws the spacecraft in, then sends it flying in a new direction. The craft moves even faster than it was traveling before, like a slingshot!

The key to gravity assists is angular momentum. This is the momentum of an object moving in an ellipse, such as an orbit. Physicists calculate a planet's angular momentum by multiplying the planet's momentum, or the speed at which it is orbiting, by the planet's distance from the sun. If Earth were moving at the same orbital speed but was twice as far from the sun, its angular momentum would be twice as great.

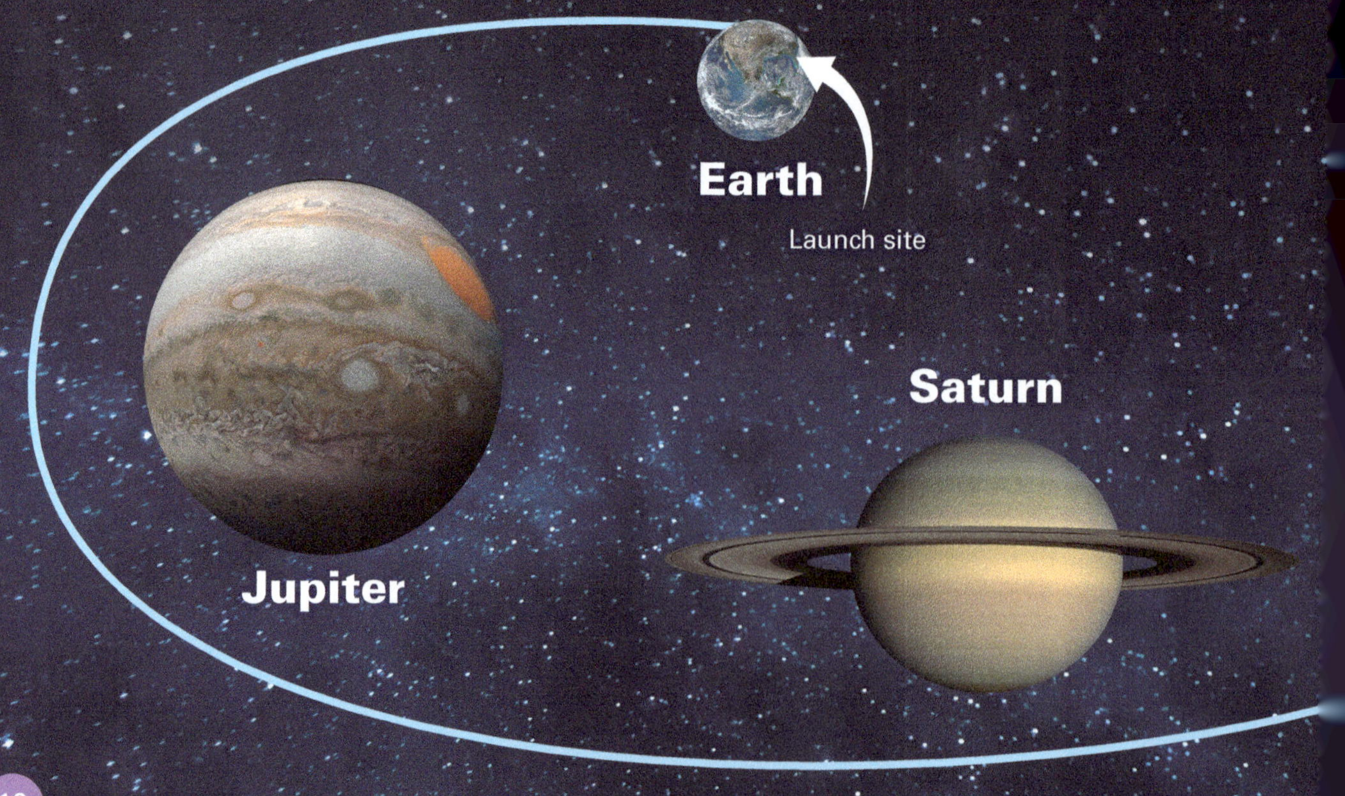

Earth

Launch site

Saturn

Jupiter

Humans have long dreamed of visiting faraway planets. But it takes a lot of energy to reach these planets! Luckily, scientists have found a clever way to use gravity to give a spacecraft a boost. It's called a gravity assist.

When a spacecraft swings past a planet, some of the planet's angular momentum transfers to the spacecraft. This has almost no effect on the planet. But it gives the tiny spacecraft a big energy boost.

Neptune

Pluto says, "Hi!"

Uranus

Coming at you, Neptune!

TECH TIME

The *Voyager 2* robotic spacecraft launched on Aug. 20, 1977. Its rocket had only enough fuel to reach Jupiter. But gravity assists at Jupiter, Saturn, and Uranus boosted the spacecraft outward into the solar system. In 2018, the tiny craft officially left our solar system, becoming one of just two human-made objects to enter interstellar space.

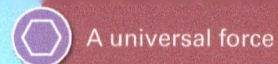

Free falling

Now imagine that the ground beneath you disappeared. Earth's gravity would still tug you downward, but without normal force pushing back, you wouldn't be able to feel it. You would be in a state known as free fall. This would make you feel weightless!

Astronauts traveling on a spacecraft, such as the International Space Station (ISS), experience a phenomenon called microgravity. This happens when an object is in free fall. As the ISS speeds through its orbit, Earth's gravity pulls the space station and everything in it downward. But with no normal force pushing back, everything inside the ISS, including its astronauts, experiences weightlessness.

From the moment you were born, Earth's gravity has been tugging on you. But gravity isn't the only *force* at work. As gravity pulls you downward, the ground beneath you pushes back against your feet. This is known as normal force. It allows you to feel the effects of Earth's gravity.

It's leg day!

No normal force? No problem!

Weightlessness is hard on the human body.
Without gravity pulling against them, bones weaken and muscles shrink.

TECH TIME

Scientists are working to develop artificial gravity. This would allow astronauts to spend more time in space. One way to do this would be to generate *centripetal force*, which causes a spacecraft to rotate. Humans inside the spacecraft would feel a gravitylike force pulling them toward the craft's outside walls. One day, millions of people could live in gigantic cylinder-shaped space stations!

ALL ABOUT ORBITS

An orbit is the elliptical, or curved, path that an object follows as it travels around a body in space. The moon orbits Earth. Earth orbits the sun. And the sun orbits the center of the Milky Way galaxy.

Everything in the universe is on the move. Even when you are sitting still, you are speeding through the solar system at 67,000 miles (108,000 km) per hour. This is the speed at which Earth orbits the sun.

I'm the slowest planet in the solar system!

An object, such as a planet, needs a lot of speed to stay in orbit. Gravity constantly tugs the planet toward the sun. This means the planet must move fast enough that it never actually hits the body it is orbiting. The planet remains in a constant state of free fall!

The spiral arms of our Milky Way galaxy swirl around a massive black hole at the galaxy's center.

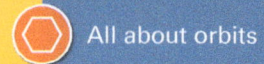

Orbiting **Earth**

Geostationary Operational
Environmental Satellites (GOES)
22,233 miles (35,780 km)

High Earth Orbit (HEO)
22,233 miles+ (35,780 km+)

**MEO
is commonly used
for navigation satellites**,
such as the ones in the *Global Positioning
System* (GPS). These satellites have a
predictable orbit that passes over the
same point twice per day.

Medium Earth Orbit (MEO)
1,243–22,233 miles (2,000–35,780 km)

**LEO is
the most common
type of orbit.** It is especially
useful for satellites that take pictures
of Earth. LEO is where most human
spaceflight happens.

Low Earth Orbit (LEO)
112–1,243 miles (180–2,000 km)

As you read this book, more than 8,000 human-made objects are orbiting Earth. They include everything from communications satellites to space stations! Orbits around Earth are divided into three main categories. Each orbit is useful for different kinds of spacecraft.

HEO is useful for weather and communications satellites because they can see the whole planet at once. When a spacecraft always stays over the same point on Earth's surface, it has a geostationary orbit.

GPS satellite
12,550 miles (20,200 km)

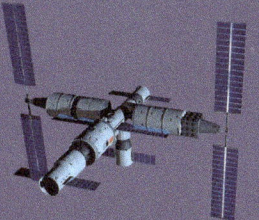

Tiangong Space Station
250 miles (402 km)

CAREER CORNER

Astronauts do much more than flip around and float in space stations. ISS and Tiangong astronauts spend much of their time doing chores and maintaining the space station. They must be house cleaners, plumbers, electricians, and IT specialists! But physics is also part of an astronaut's work. Astronauts conduct many science experiments that are possible only in microgravity.

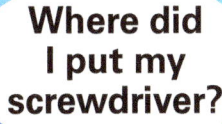

Where did I put my screwdriver?

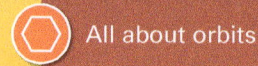

Orbits at **work**

The set of rules for how orbiting objects move is called orbital mechanics. Astronauts and engineers must follow these rules when they want to change a spacecraft's orbit.

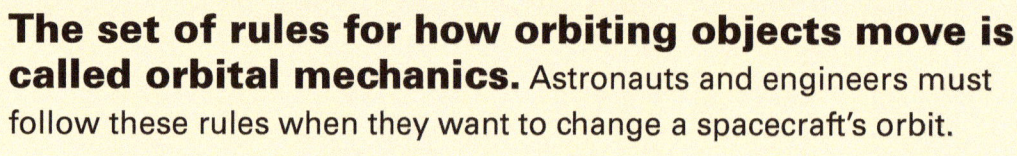

Objects in higher orbits move at slower speeds.

But going into a higher orbit requires more energy. To achieve this, a spacecraft fires its thrusters in the opposite direction to the way it's traveling. This increases the spacecraft's *tangential speed*. It pushes the spacecraft into a higher orbit, where it moves more slowly. It must speed up to slow down!

I'm outta here!

To reach orbit, a rocket must first escape the powerful pull of Earth's gravity.

If a spacecraft speeds up until it reaches its escape velocity, it leaves the planet's orbit entirely.

At some point, every single satellite, spacecraft, and space station orbiting Earth was blasted into space. But what happens when we need to change an object's orbital position?

Going into a lower orbit requires the spacecraft to lose energy. It fires its thrusters in the same direction it is traveling. This reduces the spacecraft's tangential speed. It moves into a lower orbit, where it moves more quickly. It slows down to speed up!

If a spacecraft slows down past a certain point, its orbit gets so low that gravity pulls the craft down to the planet's surface.

Broken satellites, old telescopes, and even lost nuts and bolts orbit Earth. This debris is known as space junk!

DID YOU KNOW?

In 1965, U.S. astronaut Ed White lost a rubber glove during a spacewalk.

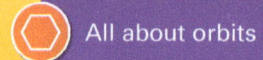

Kepler and **orbits**

German scientist **Johannes Kepler** spent years studying the motion of the planets. In the 1600's, he used physics to develop three laws describing how planetary orbits worked.

Kepler's First Law states each planet's orbit is in the shape of an ellipse. The sun is located at one *focus* of the ellipse.

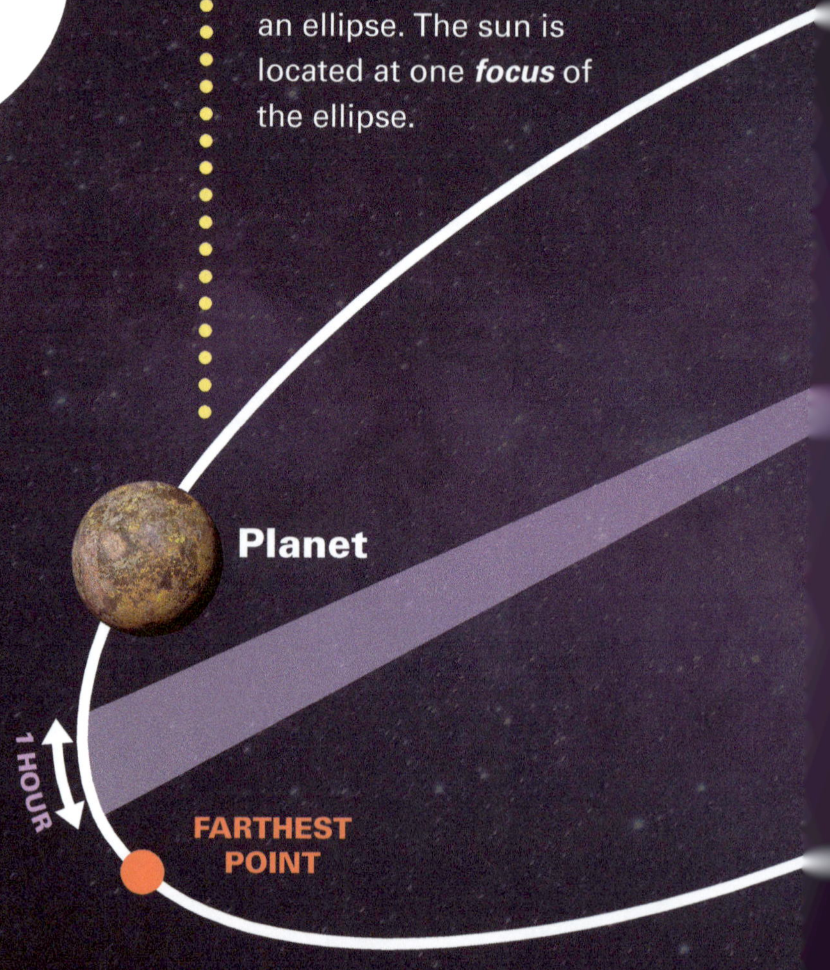

Planet

1 HOUR

FARTHEST POINT

Throughout history, humans have worked to understand how the universe works. But more than 400 years ago, Johannes Kepler solved the mystery of orbits.

Sun

CLOSEST POINT

1 HOUR

Kepler's Second Law states an orbiting planet always sweeps out the same area in a given amount of time. When moving fast and close to the sun, the planet travels a long way in one hour. The area covered is short and wide. When moving slowly and far from the sun, the planet travels a shorter distance in an hour. The area covered is tall and narrow. But the two areas will always be equal in size.

I win!

Kepler's Third Law states it takes distant planets much longer than close planets to orbit the sun. Earth takes 365 days to orbit the sun. Saturn takes 10,759 days!

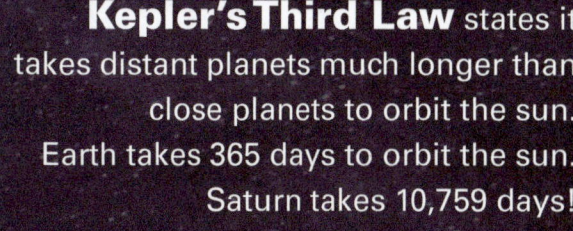

OUR AMAZING SOLAR SYSTEM

The solar system is like our cosmic neighborhood. And the sun is at the center of it all.

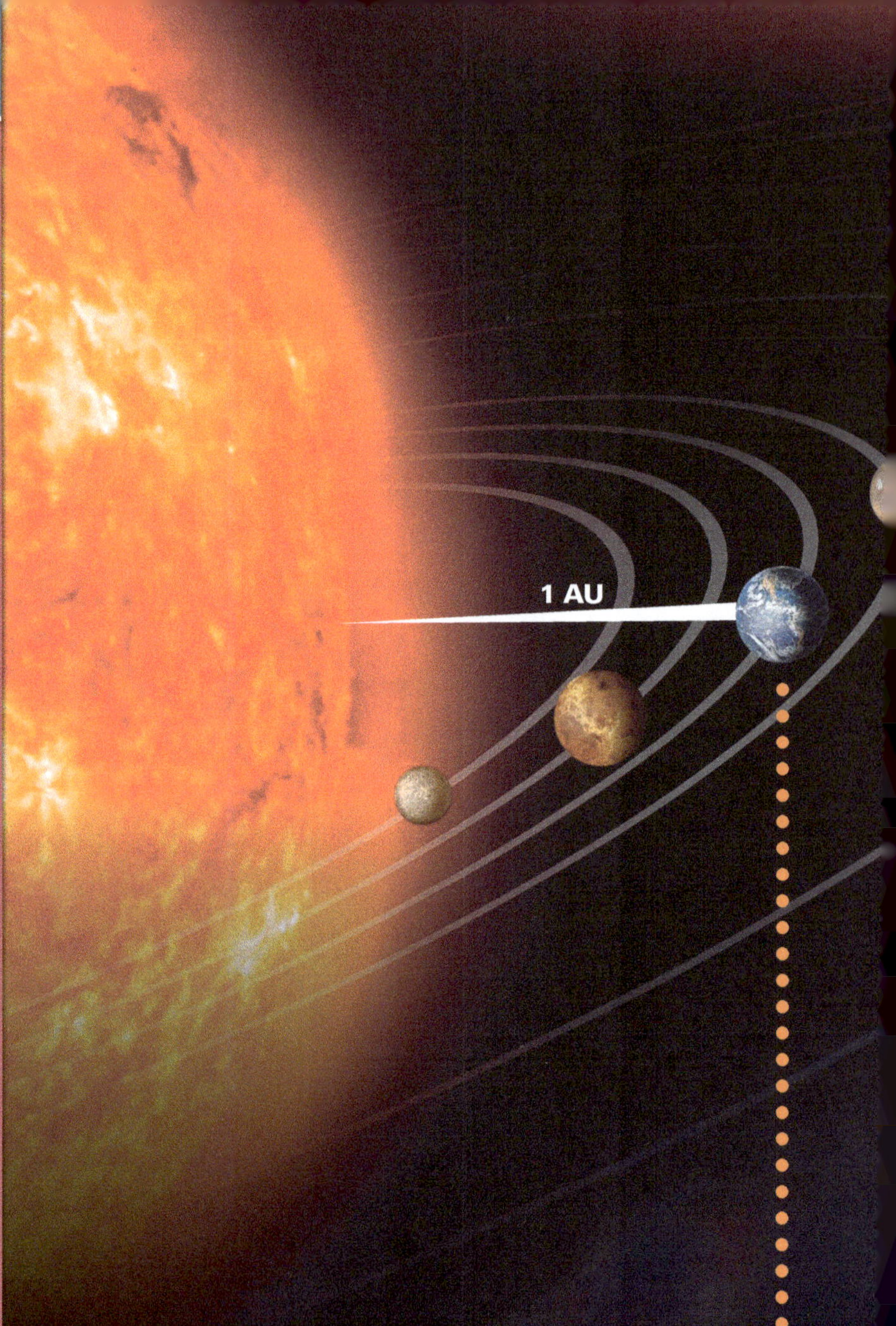

1 AU

Distances in the solar system are vast. Objects are millions or even billions of miles apart. To make talking about these distances easier, scientists use astronomical units (AU). One AU is the distance between Earth and the sun, or about 93 million miles (150 million km).

Rings are my thing!

Our solar system's known occupants:

 **Planets**
8

 *Dwarf Planets*
5

 **Moons**
290

 Asteroids
1,300,000+

 **Comets**
3,900+

What else?

Scientists are still discovering new objects in our solar system. Some think there may be a planet X orbiting past Neptune, just waiting to be discovered!

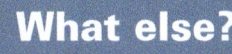

Your home **star**

The sun has an enormous amount of mass. In fact, the sun makes up about 99.8 percent of all mass in the solar system! Planets, moons, asteroids, and other objects make up the other 0.2 percent.

German physicist **Albert Einstein** formalized the equation $E = mc^2$ in 1905. The equation means that energy (E) equals mass (m) times the speed of light (c) squared. It shows that a small amount of mass can be turned into a lot of energy.

Fusion

The core's heat and pressure are so extreme that hydrogen atoms' *nuclei* fuse to form helium nuclei. During this fusion process, the nuclei lose a small amount of mass.

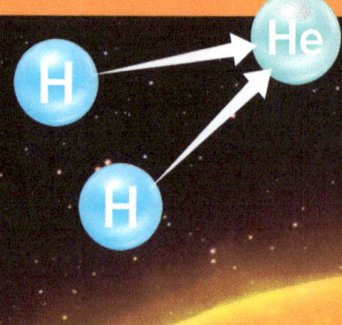

Without our sun, we wouldn't have a solar system. The sun's heat and light make life on Earth possible. And the sun's gravity binds all the solar system's objects together.

The sun's huge mass means it has a strong gravitational pull. The gravity pulls inward, producing extreme heat and pressure at the sun's core.

Mass and energy

The tiny amount of mass lost during the fusion process is converted into a huge amount of energy.

DID YOU KNOW?

The sun's core is so dense that one cup of it would weigh about 80 pounds (36 kg)!

Light

This energy travels outward through the sun's layers. It takes about 170,000 years to reach the surface. From there, it spreads out in all directions as light.

Measuring space with light

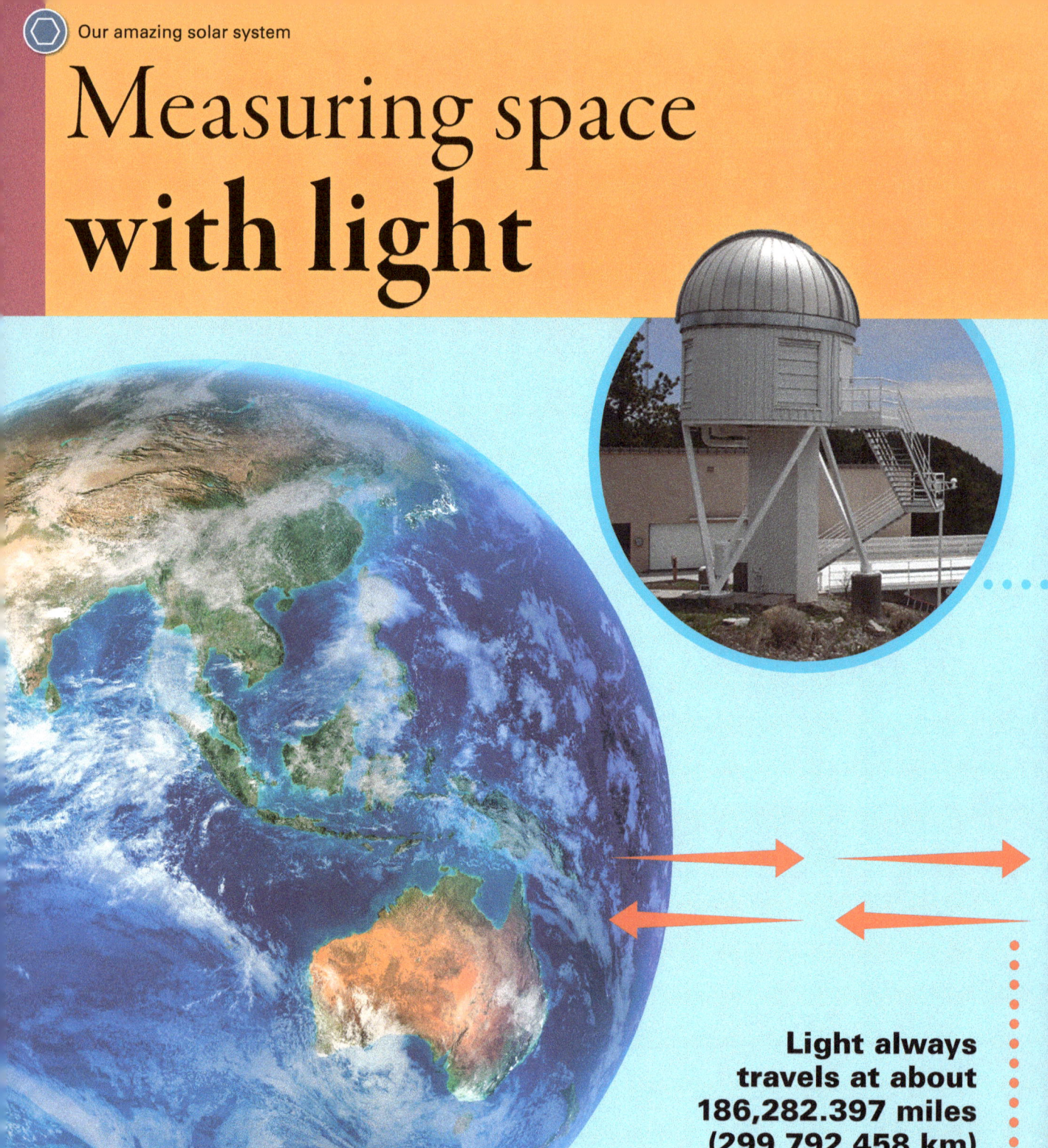

Light always travels at about 186,282.397 miles (299,792.458 km) per second.

Scientists use this fact to measure the vast distances in space.

Sunlight is made of tiny packets of energy called photons. And these photons are *fast!* In fact, they are the fastest things in the universe. As far as we know, nothing can move faster. This makes the speed of light the universal speed limit!

In the 1960's and 1970's, astronauts placed devices called retroreflectors on the moon's surface. A retroreflector bounces light back in the exact direction it came from.

Scientists fired laser pulses at the moon from an observatory on Earth. They timed how long the pulses took to bounce back. This let them figure out the precise distance to the moon at any given time.

Bye-bye, Earth!

DID YOU KNOW?

The fastest human-made object is the Parker Solar Probe. In 2023, the probe hit about 0.06 percent of the speed of light!

The measurements showed that the moon is traveling away from Earth. It moves about 1.5 inches (3.8 cm) farther away each year!

Crashing into asteroids

Most asteroids orbit in the Asteroid Belt, a region of our solar system that lies between Mars and Jupiter. But sometimes an asteroid's orbital path takes it dangerously close to Earth. In 2013, a fireball exploded in the Russian sky. The blast generated a huge wave of energy. For the first time in more than one hundred years, a large asteroid had struck Earth. Scientists are exploring ways to use physics to change the path of dangerous asteroids.

ORIGINAL ORBIT

Didymos

Dimorphos

In 2022, scientists crashed a spacecraft into a nearby asteroid called Dimorphos. This small asteroid orbited a larger asteroid called Didymos.

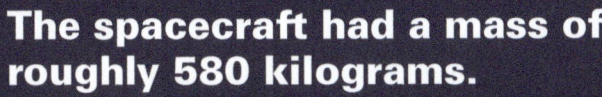

Asteroids are rocky, metallic, or icy bodies left over from the solar system's formation.

The spacecraft had a mass of roughly 580 kilograms.

Dimorphos's mass was estimated at 5 billion kilograms. Still, the tiny spacecraft transferred enough energy to Dimorphos to change the asteroid's orbit! The mission proved scientists could someday use a similar mission to stop an asteroid from hitting Earth.

NEW ORBIT

Uh-oh!

 # CURIOUS CONNECTIONS

GEOLOGY

About 66 million years ago, 75 percent of Earth's animals went extinct, including the dinosaurs. In the 1970's, a team of geologists noticed something strange about ancient rock layers they were studying. The rocks contained high amounts of the element iridium. This element is very rare on Earth, but it is common in asteroids. The geologists hypothesized that an asteroid had hit Earth millions of years ago. Today, most scientists blame an asteroid for the dinosaurs' demise.

ROCKET POWER

For a spacecraft to reach outer space, it must first overcome the mighty pull of Earth's gravity. This is no easy task. The spacecraft must create enough upward *force* to overcome Earth's gravity. To achieve blastoff, a spacecraft uses a rocket.

Outer space is my place!

Rockets are powered by propellants. Under the right conditions, these substances produce hot, high-speed gases that can push a rocket forward. However, propellants are heavy. The more propellant a rocket carries, the more propellant it must burn to achieve thrust.

Building a rocket in multiple stages makes it easier to launch into space. The first part of the rocket, or stage, uses up its propellant. Then it falls away. The rest of the rocket has its own engines and propellant. It doesn't need to carry the weight of the empty first stage.

Anyone going to Mars?

The spacecraft *Psyche* blasts off for its mission to study an asteroid.

3-2-1 blastoff!

Earth is a challenging place to launch a rocket. Understanding physics can mean the difference between a rocket successfully blasting into space or crashing back to the planet's surface. Engineers must consider four major forces during a rocket launch.

Lift

Lift pushes perpendicular to the direction of travel. It is caused by air flowing over the rocket. Lift helps keep the rocket stable in flight.

Drag

Drag pushes opposite the direction of travel. The rocket must push through any air molecules, slowing it down.

Thrust

Thrust pushes a rocket forward. It is created when propellants exit the back of the engine at a high speed.

Gravity

Gravity pulls the rocket toward the ground. The strength of this force depends on the mass of the planet or moon from which the rocket is launched.

The *forces* in a rocket launch are affected by the gravity and atmosphere of the launch location. Stronger gravity creates a more powerful downward force on the rocket. A thicker atmosphere creates more drag and lift. Rocket engines can generate more thrust in a thinner atmosphere.

Earth

Mars

Earth's Moon

Atmosphere
Thick

Gravity
1 g

Atmosphere
Thin

Gravity
0.38 g

Atmosphere
Almost none

Gravity
0.17 g

Isaac Newton was a physicist and mathematician who lived in the late 1600's. He created a set of rules that described the relationship between forces and motion on an object. These rules are known as Newton's laws of motion. Rocket *propulsion* can be explained by Isaac Newton's third law of motion. It states that every action has an equal and opposite reaction.

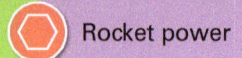

Rockets and forces

Solid rocket engines

carry a premixed fuel-*oxidizer* propellant. The mixture is ignited to create *combustion*.

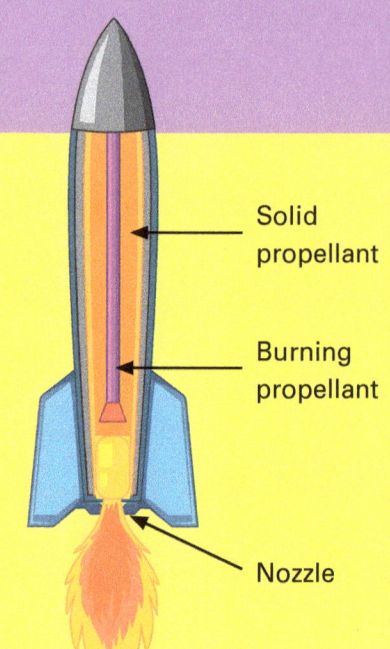

Solid propellant

Burning propellant

Nozzle

Liquid rocket engines

use two separate propellants. One is a fuel, such as liquid hydrogen. The other is an oxidizer, such as liquid oxygen. When the propellants are mixed and ignited, combustion occurs.

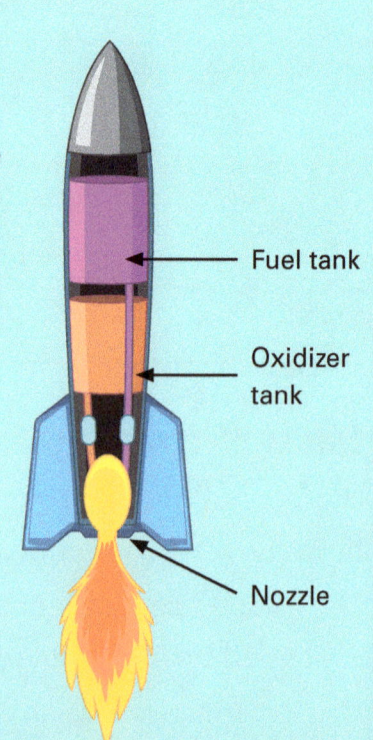

Fuel tank

Oxidizer tank

Nozzle

Have you ever watched a rocket launch in a movie or TV show? You probably noticed bright white-orange flames streaming out the back of the rocket as it lifted off the launchpad. These flames are actually pressurized gas. Read on to learn how different kinds of rockets achieve liftoff!

- *Ion* **rocket engines** carry a gas, such as xenon. A device bombards the atoms with electrons, creating positively charged ions. These ions are accelerated out the back of the engine, which propels the rocket forward.

Ion engines emit a blue glow.

- **Ion engines harness the power of atoms** to achieve thrust. All atoms contain three types of subatomic particles. Protons carry a positive electric charge. Neutrons carry a neutral charge. Electrons carry a negative electric charge.

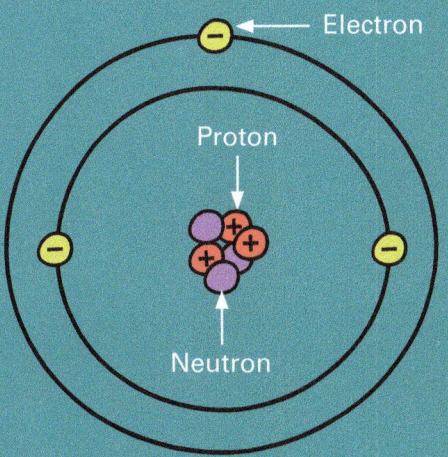

Electron

Proton

Neutron

CAREER CORNER

Do you love math, science, and all things STEM? Then you might be a future rocket scientist! Rocket scientists are a type of aerospace engineer. Their job is to design, build, and test rockets for spacecraft and satellites. Many rocket scientists work for space programs, designing probes and space capsules.

Recycling rockets

Launch
The Falcon 9 is a two-stage rocket. Its first stage has nine liquid-fueled engines.

Ascent
The rocket angles to the side so it can gain the speed it needs to reach orbit.

Stage Separation
The first stage separates from the second stage.

You launch here too?

For most of the history of spaceflight, rockets were used just once and then thrown away. The lower stages of a rocket crashed into the ocean while the rest of the spacecraft continued into space. But in 2017, the company SpaceX launched the first-ever recycled rocket, the Falcon 9.

Second Stage
The second stage continues to orbit.

Flip and Reentry
The first stage flips around and fires three engines to decelerate. It needs to slow down because reentering the atmosphere at high speed generates intense heat.

Steering
Grid fins at the top of the stage help steer toward the landing site.

Landing
The first stage fires one engine to bring it to a soft touchdown on a landing ship at sea.

CURIOUS CONNECTIONS

MATH

Landing a rocket requires a lot of math. Engineers must figure out how to steer and land a falling rocket stage safely while using as little fuel as possible. This saves fuel for launching bigger *payloads* into space. Experts at SpaceX published mathematics papers covering new ways to do this. Their work helped make reusable rockets possible.

DEEP SPACE

Our solar system is just one of thousands of star systems in the Milky Way galaxy. Scientists believe that beyond the Milky Way, there could be trillions of other galaxies. Some may date back to the beginning of the universe itself. And sprinkled across galaxies lie black holes, *quasars*, *nebulae*, and other deep space mysteries just waiting to be solved.

Scientists believe a supermassive black hole lies at the center of every galaxy.

The universe is a big place. Right now, the moon is the only other celestial body humans have set foot on. But that doesn't stop scientists from dreaming about launching missions to explore the farthest reaches of deep space.

I'm a star baby!

Newborn stars form in pillars of dust and gas inside the Eagle Nebula.

The space between stars, planets, and other celestial objects is a vacuum.

It is empty of *matter*. However, space is not a perfect vacuum. Even parts of deep space that appear to be completely empty have atoms floating around. The farther you get from large objects in space, the emptier the vacuum is.

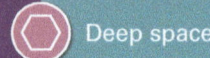

Exoplanet
exploration

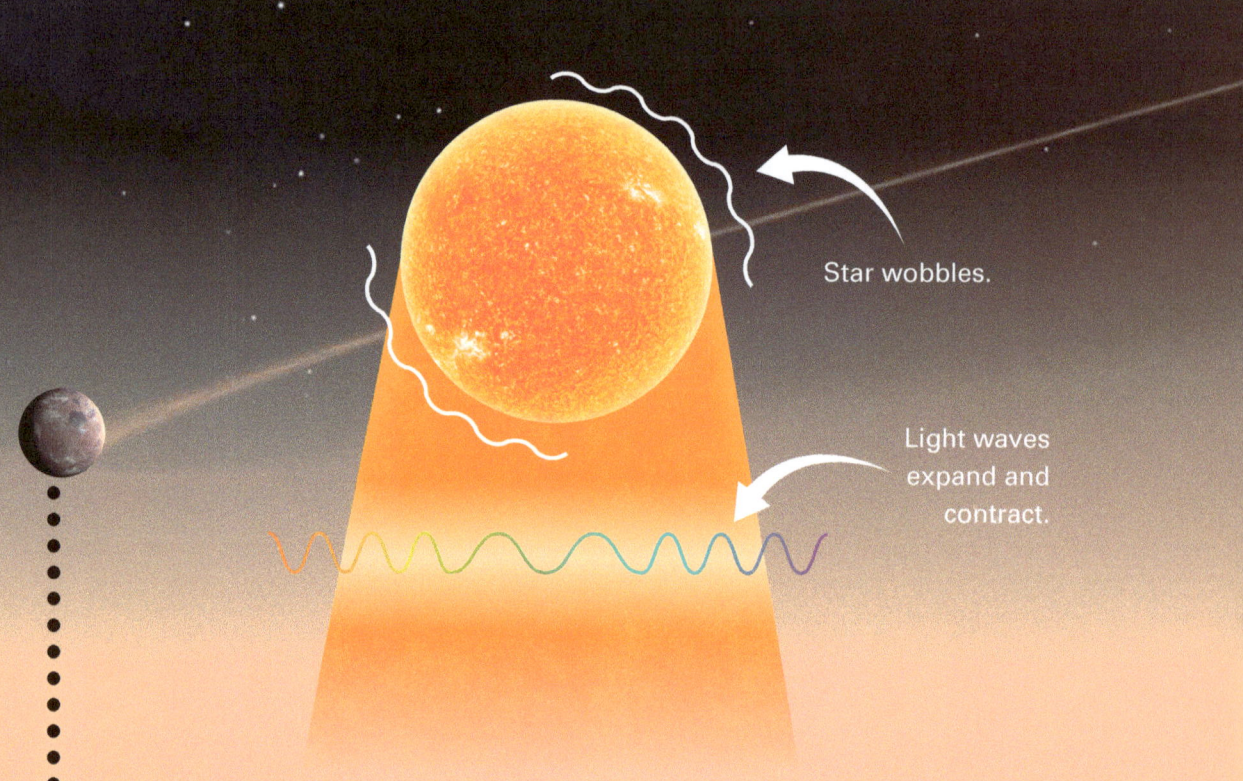

Star wobbles.

Light waves
expand and
contract.

**One way scientists search for exoplanets is
called the radial velocity method.** As a planet orbits
a star, the planet's gravitational pull causes the star to wobble
slightly. As the wobble moves the star toward Earth, its light waves
compress. As the wobble moves the star away from Earth, its light
waves expand. This causes the starlight's color to change slightly.
These tiny color shifts tell scientists that a star has planets, how
many planets it has, and even how large those planets are.

Exoplanets are planets outside our solar system. Some may be covered in icy oceans or active volcanoes. Others may be home to alien life forms, just waiting to be discovered!

In the transit method, scientists look for the dimming of a star's light.

This happens when a planet transits, or passes in front of, a star. The amount of dimming provides clues about how large the planet is and how far away it is from its star. Analyzing the colors in the light can tell scientists about the star's atmosphere. They might even provide clues about possible life on those planets!

The star's light dims slightly.

Venus transiting the sun

DID YOU KNOW?

Scientists have discovered more than 5,000 exoplanets!

Don't mind me!

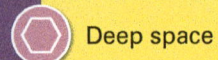

The death of a **star**

Medium-sized stars like our sun cool gradually, eventually collapsing into a dim, Earth-sized star called a white dwarf. Finally, the star cools completely, becoming a black dwarf.

I used to be yellow!

When massive stars run out of fuel, the intense gravitational pull of the core causes the star to collapse in on itself, forming a densely packed core. The intense pressure and heat energy inside the core build and build. Eventually, the *matter* and energy from the star's outer layers release in a massive explosion called a supernova.

DID YOU KNOW?

One teaspoon of a neutron star would weigh 10 million tons (9.1 million metric tons)!

No star lives forever. A star's core contains a finite amount of the hydrogen atoms needed for fusion. Eventually, this hydrogen fuel runs out.

PULSAR

MAGNETAR

After the supernova, the remains of the star's core may form a neutron star. Neutron stars are only about 12.5 miles (20 km) across. But they are the densest objects scientists have discovered. There are two types of neutron stars: pulsars and magnetars. Pulsars spin rapidly and give off powerful beams of light. Magnetars have extremely powerful magnetic fields. These magnetic fields trigger huge explosions of energy that can be detected from Earth.

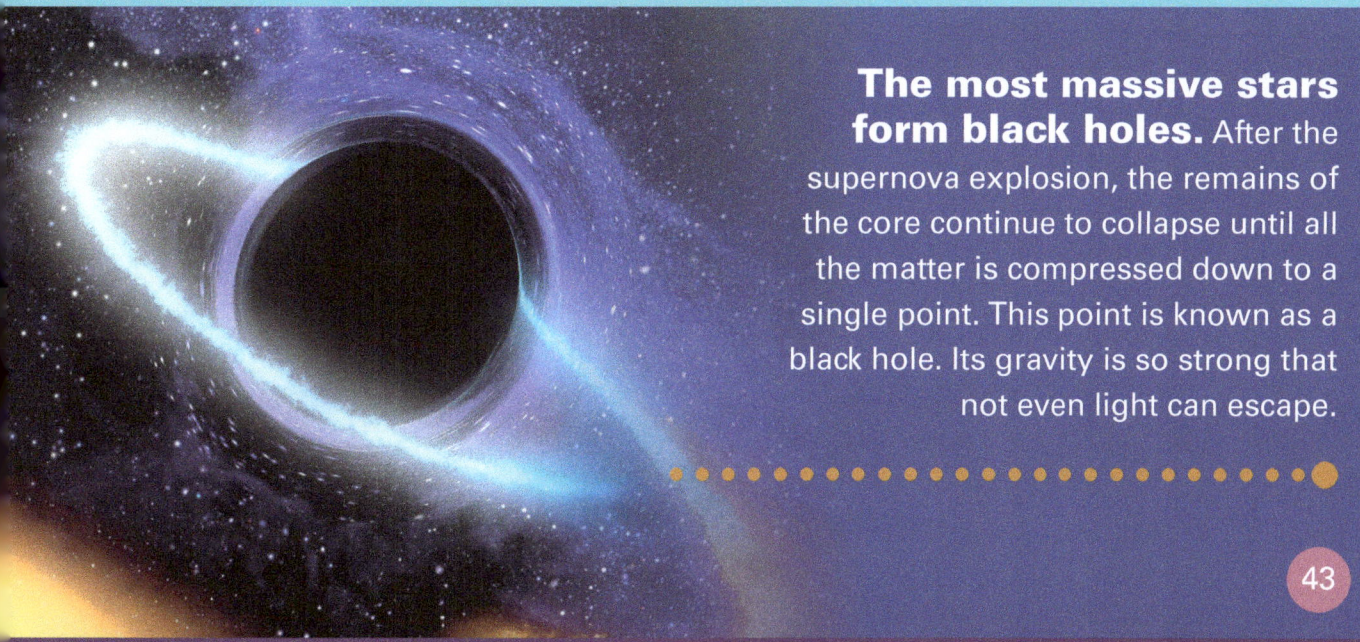

The most massive stars form black holes. After the supernova explosion, the remains of the core continue to collapse until all the matter is compressed down to a single point. This point is known as a black hole. Its gravity is so strong that not even light can escape.

Escaping gravity

You will need:

- Water balloons filled with various amounts of water (e.g., 50 grams, 100 g, 200 g)
- 5 feet (1.5 m) of string
- Clothespin (the type with a metal spring)
- A large outdoor space with about 20 feet (6 m) of clear space in all directions

Give it a try

1. Attach the string to the clothespin by threading it through the metal spring and tying a double knot. The clothespin should hang from the end of the string.
2. Carefully clip the filled water balloon to the clothespin.
3. In an open outdoor area, hold the end of the string and slowly swing the water balloon around yourself. Spin just fast enough to keep the balloon a few feet above the ground.
4. Slowly increase your spinning speed until your water balloon "escapes" gravity.

Try this next!

Repeat this process with water balloons of different sizes and weights. Pay careful attention to what causes the water balloon to leave its orbit and how it moves.

Gravity is an important *force* in our lives. It keeps us grounded on Earth and keeps celestial bodies orbiting in our solar system. But how do astronauts escape the pull of gravity so they can explore space? And how does an orbiting object, such as the ISS, stay in its orbit?

QUESTION TIME!

How does this activity relate to what you learned from this book about gravity in space and space travel? What does it take for a celestial object to escape its orbit?

Index

Glossary

centripetal force (sen TRIP uh tel FORS)—an inward force experienced by an object that keeps it traveling around another object

combustion (kuhm BUHS chuhn)—a chemical reaction that produces heat and light

dwarf planet (DWORF PLAN iht)—a round body that orbits the sun but whose gravity is not strong enough to clear other objects from its orbit

focus (FO kuhs)—one of two stationary reference points that help define an ellipse

force (FORS)—energy acting on an object

Global Positioning System (GLOH buhl puh ZIH shuhn ing SIH stuhm)—a group of satellites operated by the U.S. government that beam signals to Earth; these signals allow devices on Earth to determine the devices' exact location

ion (AI on)—a particle, such as an atom, that has lost or gained electrons

matter (MAT uhr)—the material that makes up physical objects

nebula (NEB yoo luh)—a cloud of dust and gas where new stars may be born

nucleus (NU kli uhs)—the center, or core, of an atom

oxidizer (OX ih dai zuhr)—a substance that releases oxygen atoms, causing a chemical reaction

payload (PEI lod)—the load a spacecraft carries, such as people or fuel

propulsion (pruh PUHL shuhn)—the act of moving an object forward

quasar (KWAY zayr)—the superbright center of a galaxy that is extremely far away, likely produced by an active black hole

tangential speed (tan JEN shuhl SPEED)—the speed of an object moving in a circular path

www.ingramcontent.com/pod-product-compliance
Lightning Source LLC
Chambersburg PA
CBHW052142170526
45159CB00017B/3138

* 9 7 8 0 7 1 6 6 7 1 5 5 8 *